How Animals Survive

Written by Barbara Donovan

Table of Contents

Food and Water 4
Safety 8
Shelter 12

What do animals need to stay alive?

They need the same things you do! They need food when they get hungry. They need water when they get thirsty.

They need to keep safe and to find a shelter that protects them from the weather and other animals.

Food and Water

A jaguar's good eyesight helps it hunt at night.

You feed your pets, so they don't have to find food and water. But wild animals do. They use sight, sound, and smell to help them find food to eat and water to drink.

A polar bear can smell a seal hiding under the ice.

Did you know that some animals work together to find food? Groups of dolphins hunt fish by working together to round them up.

Dolphins

Wolves hunt in groups called packs.

Wolf pack

Geese

In very cold, snowy weather, it's hard for animals to find enough food and water. So many animals move, or **migrate**, to warmer places where there is more to eat.

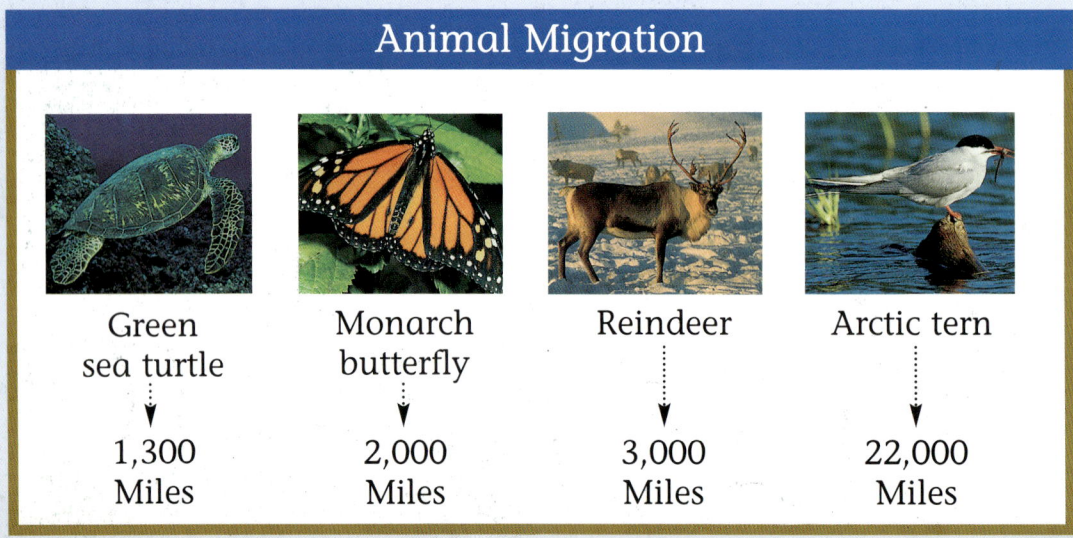

Animal Migration

Green sea turtle	Monarch butterfly	Reindeer	Arctic tern
1,300 Miles	2,000 Miles	3,000 Miles	22,000 Miles

It's also hard for animals to find enough food and water during very hot weather. Water dries up and plants die. In Africa, thousands of wildebeests migrate when the dry season comes.

Wildebeests

Safety

No animal can outrun a cheetah, the fastest land animal.

Animals have many different ways of dealing with **predators**. Predators are animals that eat other animals for food.

Some animals hide from predators. Others run fast to get away.

Rabbit

Animals may signal or **communicate** with each other to warn that a predator is near. Meerkats peep, growl, or bark to warn of danger. As danger gets closer, their cries get louder.

Meerkats

Turtles have hard shells that keep them safe. Skunks have a spray that warns, "Stay away!" Porcupines have long, sharp quills that help protect them from harm.

Skunk

A hawk lands on the hard shell of a giant tortoise.

When zebras stand close together, it's hard to pick out just one animal. This confuses their predators.

Camouflage protects some animals by helping them blend in with their surroundings. Octopuses can change their color and patterns so that predators cannot see them.

Octopus

Shelter

Red fox kits

Your home is a **shelter** for you and your family. Animals need shelter, too. An animal's home protects it from bad weather. It also keeps an animal safe while it eats and sleeps.

Mother polar bears dig a den in the snow.

Some animal shelters are high up in trees. Orangutans build nests each night out of leaves and branches. Some of the nests are 100 feet above the ground.

Orangutan nest

Panda

Raccoon

Hollow trees also make good shelters for animals. Mother pandas and raccoons give birth in hollow trees.

Some animals need shelter when they **hibernate**, or spend the winter sleeping. Animals that hibernate eat little or no food during the cold winter months.

A hibernating woodchuck loses up to one half its weight.

A cave is a good place for a bear to hibernate.

15

Squirrel monkey

It's not easy for an animal to stay alive in the wild. Just think of the many ways animals find food, water, safety, and shelter.